CHEM TUTOR CD-ROM

Welcome to Chem Tutor, the step-by-step multimedia guide to the essentials of introductory college chemistry!

INSTRUCTIONS

To install Chem Tutor, you need 18 MB free space on your C hard drive.

WINDOWS 95 USERS

INSTALL:

    (1) Put the Chem Tutor CD-ROM disk in the drive.

    (2) Click RUN

    (3) At the prompt, type

        X:\SETUP.BAT

    where X is the drive letter of your CD-ROM (any letter from D to L is supported).

RUN:

    (1) Click START -then- PROGRAMS

    (2) In the Chem Tutor folder, click on the Chem Tutor icon.

Note: The Windows 95 taskbar is not compatible with Chem Tutor. It should therefore not be set to "always on top" or "auto hide". To be sure it is not, right click on the taskbar and choose "properties". If for any reason the taskbar appears while Chem Tutor is running, do not click on it.

WINDOWS 3.1X USERS

INSTALL:

    (1) Use the Run command from the File Menu in the Program Manager.

    (2) Type X:\SETUP.EXE in the space provided, where X is the drive letter of your CD-ROM.

RUN:

Open the Chem Tutor folder and double click on the Chem Tutor icon.

CHEM TUTOR STUDENT WORKBOOK AND CD-ROM

JOHN D. LAMB Ph.D.
Brigham Young University, Utah

Jones and Bartlett Publishers

Sudbury, Massachusetts

Boston   London   Singapore

Editorial, Sales, and Customer Service Offices
Jones and Bartlett Publishers
40 Tall Pine Drive
Sudbury, MA 01776
info@jbpub.com
http://www.jbpub.com

Jones and Bartlett Publishers International
Barb House, Barb Mews
London W6 7PA
UK

ISBN 0-7637-0350-8

Printed in the United States of America
00 99 98 97 96   10 9 8 7 6 5 4 3 2 1

Preface

So you want to learn chemistry? Some students suffer a shock when they first encounter college or advanced high school chemistry–it has been compared to drinking from a firehose. This tutorial is designed to help you achieve your lofty learning goal. In fact, I would like to think that it might even make the process rather enjoyable.

Well over a decade of experience in teaching introductory college chemistry has been distilled into this work. And "distilled" is a good term for reasons besides the obvious pun. This tutorial is really concentrated. If you have a chemistry text (as you should), you have noticed that it contains a great deal of information, much more than you could hope to learn. This is good–some of it makes for interesting reading, some covers topics in depth for the truly curious, some provides exhaustive exercises to hone your problem-solving skills, some is designed to make the book really heavy so you feel like you got your money's worth. For these reasons and more, your textbook is invaluable. But in this tutorial, I have tried to distill out only the essential elements (another pun!) of the introductory chemistry course.

Any daunting task like learning college chemistry is made easier by breaking it down (decomposing it) into its component parts. The material in this course can be decomposed into three parts:

(1) *Chemical Vocabulary: names of concepts and chemical species.* Commit to memory the jargon of chemistry. So much of learning a discipline like chemistry is really like learning a new language. If you can communicate with ease (both speaking and hearing) in this language, it will make all the rest much simpler.

(2) *Chemical Concepts: laws, principles and structures.* Much of the tutorial is dedicated to explaining these concepts, but it is up to you to commit them to memory and internalize their meaning and ramifications. Look to your teacher and text for more in-depth explanations where needed.

(3) *Chemical Problem Solving.* You will learn this best from doing the practice problems–along with additional problems from your textbook.

This workbook is designed to help you learn these three kinds of skills from the information in the tutorial. *Use it!* Each Chapter is divided into three sections: Vocabulary, Concepts, and Problem Solving. In the Vocabulary section, write the definition for each chemical term you encounter (underlined in the Tutorial). In the

Concepts section, make notes on each new concept you encounter. In the Problem
Solving section, try each practice problem in the workbook before looking at the step-by-
step solution on the computer. Your problem solving muscles will develop faster if you
struggle with the problem before seeing the solution.

Some Words of Wisdom

Practice! The more you do, the easier chemistry will get. In this respect, learning
chemistry is like learning to play the piano or a sport. You should treat your study time as
if it were practice time. Practice chemistry the same way you would learn to shoot 3-
pointers in basketball.

 The knowledge you will gain in chemistry is cumulative. The understanding of each new
concept will depend on a mastery of the previous concepts. So it is best to take the topics
in sequence if you are seeing the material for the first time.

How can this tutorial help me be a chemistry superstar (or at least pass the course)?

ChemTutor is designed to walk you step by step through the essentials of introductory
chemistry. It can be used as a first-time learning device, especially if used in conjunction
with one of the many excellent chemistry texts available. If you are currently taking a
chemistry course in school, it can help you prepare for exams; or it can help you cover
material you missed when you were sick or when you were on that Geology field trip. It
is an ideal tool for reviewing material you once knew, for example in preparation for
medical or graduate school entrance exams. It is also great for the working professional
who needs to learn or relearn the basics of chemistry to keep or upgrade a job! Or, if you,
like Hercule Poirot, simply enjoy intellectual pursuits, it can really help to develop those
"little grey cells". ENJOY.

Acknowledgements

Thanks to the following hard-working, talented people who have assisted in the
preparation of Chem Tutor: Jared Parker, Lyneen Cluff, Stacey Staples, Kurtis Staples,
Robert DeMille, Robert Flynn, Karl Czirr, Michael Lamb, Joshua Lamb, Spencer Melby,
Jeff Silk, Morgan Greenwood, Mark Christensen, Collin Driscoll and Peggy Erickson and
the secretarial staff of the BYU Department of Chemistry and Biochemistry. Thanks to
three talented student cartoonists: Jeremy Lamb, Randy Astle, and Robert Flynn. Special
thanks to my colleague Coran Cluff for detailed proofreading and feedback. Thanks also
for support and encouragement from my Department Chairs, Earl Woolley and Fran
Nordmeyer, and from Blaine Bowman, President of Dionex Corporation. And finally,
thanks to my wife Betty for enduring all my projects.

John D. Lamb

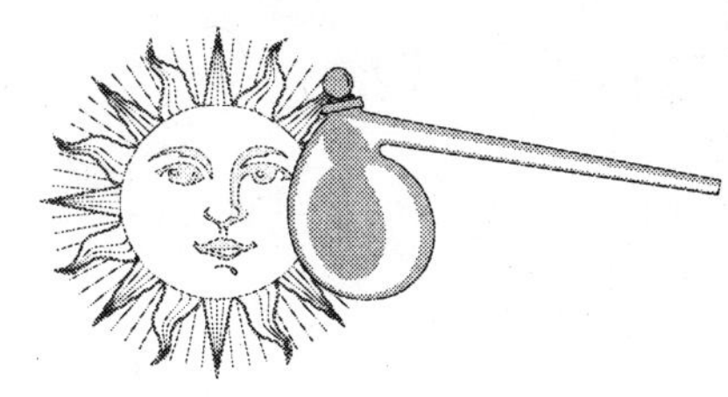

CHAPTER ONE

Basic Tools and Principles

Categories of Matter

Stoichiometry–Fundamental Principles

History, Law of Definite Proportions, Law of Conservation of Mass

Tools Part I

Significant Figures, Scientific Notation, Precision, Avogadro's Number

Tools Part II

Units, Density, Temperature Scales and Conversions

Unit Conversions and Dimensional Analysis

Problem Solving

Reactions and Stoichiometry

Compounds, Simplest Formula, Chemical Equations, Balancing Reactions

Reaction Yield

Theoretical, Actual, and Percentage Yield, Limiting Reactant

1.A.　　Vocabulary

Write your definition of each term in the space provided.

Chemistry

Heterogeneous Matter

Mixture

Homogeneous Matter

Pure Substance

Element

Compound

Binary Compound

Ternary Compound

Quaternary Compound

Covalent

Ionic

Stoichiometry

Atom

Electron

Nucleus

Proton

Neutron

Atomic Mass Unit

Molecule

Chemical Formula

Diatomic Molecule

Triatomic Molecule

Significant Figure

Scientific Notation

Precision

Accuracy

Avogadro's Number

Mole

Atomic Weight

Isotope

Density

Temperature

Kelvin

Celsius

Fahrenheit

Absolute Zero

Unit Conversion

Dimensional Analysis

Empirical Formula

Molecular Formula

Percent Composition

Molar Mass

Chemical Equation

Balanced Equation

Theoretical Yield

Actual Yield

Percentage Yield

Limiting Reactant

1.B.　　Concepts

Write your notes on each concept in the space provided.

Law of Conservation of Mass

Law of Definite Proportions

Basic Structure of Atom

Rules for Significant Figures and Rounding

SI Units and Prefixes

Temperature Scales and Conversions

Unit Conversions and Dimensional Analysis

Balancing Reactions / Equations

1.C. Problem Solving

Write your answer to each problem in the space provided. Try the problem yourself before reading the answer in the Tutorial.

1.1. In an instrument called a mass spectrometer, copper atoms are found to be 30.91% of the isotope that weighs 64.93 amu, while the other 69.09% weigh 62.93 amu. What is the average weight of natural Cu atoms?

1.2. Normal human body temperature is about 98.6° F. What is this temperature in °C? In K?

1.3.   Convert 2.32 lb to g.

1.4.   Convert 33 mi to cm.

1.5.   Balance the following equations:

(a) ___ $CuSO_4$ + ___ KI = ___ CuI + ___ I_2 + ___ K_2SO_4

(b) ___ $HSbCl_4$ + ___ H_2S = ___ Sb_2S_3 + ___ HCl

1.6. 10.0 g of butane are burned with 10.0 g of oxygen gas according to the reaction:

$$2 \; C_4H_{10} + 13 \; O_2 = 8 \; CO_2 + 10 \; H_2O$$

(a) Identify the limiting reactant.

(b) How many grams of carbon dioxide are produced?

1.D.   Other Notes

CHAPTER TWO

Periodic Table & Reactivity

The Periodic Table

Atomic Mass Units, Atomic Weight, Groups of Elements

How Elements React–Ionic

Valence Electrons, Octet Rule, Metals & Nonmetals, Naming Ions

Ionic Compounds

Formulas, Nomenclature, Polyatomic Ions

How Elements React–Covalent

Lewis Dot Structures, Multiple Bonds, Formal Charges, Resonance

Oxidation States & Covalent Compounds

Assigning Oxidation Numbers, Naming Covalent Compounds

Reaction Types

Precipitation, Redox, Acid-Base

Stoichiometry Problems

Percent Composition, Simplest and Molecular Formulas, Practice Problems

2.A.   Vocabulary

Write your definition of each term in the space provided.

Element

Atomic Number

Atomic Symbol

Molar Mass

Atomic Weight

Periodic Table

Column (Group)

Row (Period)

Metal

Nonmetal

Metalloid

Main Group Element

Transition Metal

Alkali Metal

Alkaline Earth Metal

Chalcogen

Halogen

Noble Gas

Lanthanide

Actinide

Ion

Cation

Anion

Monatomic Ion

Core Electron

Valence Electron

Octet Rule

Ionic Compound

Polyatomic Ion

Lewis Dot Structure

Multiple Bond

Lone Pair

Formal Charge

Resonance Structure

Valence Shell Expansion

Oxidation State

Dissolution Reaction

Precipitation Reaction

Aqueous

Oxidation-Reduction Reaction

Oxidized

Reduced

Neutralization Reaction

Acid

Base

2.B.　　　Concepts

Write your notes on each concept in the space provided.

Structure of the Periodic Table

Masses of Elements

Formation of Anions and Cations

Octet Rule

Formulas and Names of Ionic Compounds

Formation of Covalent Compounds

Lewis Dot Structures

Assigning Oxidation States

Naming Covalent Compounds

Types of Reactions

2.C.   Problem Solving

Write your answer to each problem in the space provided. Try the problem yourself before reading the answer in the Tutorial.

2.1.   Name the product when:

strontium reacts with nitrogen

selenium reacts with potassium

cobalt reacts with chlorine (giving Co^{+3} ion)

sulfur reacts with copper (giving cuprous ion)

2.2.   Name the following compounds:

CsF

$TiBr_2$

BaI_2

Fe_2O_3

2.3. Draw the Lewis dot structure of SO_2 .

2.4 Assign oxidation numbers to each atom in the following compounds or ions:

$CaCl_2$

Na_2S

LiH

RbO_2

H_2SO_4

PO_4^{-3}

MnO_4^-

2.5. Give unambiguous names to the following:

N_2O_3

N_2O_5

BN

OF_2

SeF_6

2.6. What do 2.500 moles of water weigh in grams to 4 significant figures?

2.7. How many atoms are in 3.00 moles of water?

2.8. What is the total number of protons associated with the chlorine atoms in 3.00 moles of carbon tetrachloride?

2.9.   A compound is 30% nitrogen and 70% oxygen. What is the simplest formula?

2.10.   Ammonium dichromate (an orange solid) decomposes with a flame to form
dichromium trioxide, nitrogen gas, and water. How many g of water are produced
if we decompose 12.0 g ammonium dichromate?

2.D.   Other Notes

CHAPTER THREE

Atomic Structure

Quantum Mechanics and Light

Light: Wave vs. Particle, Spectra, Planck's Equation, Wave Characteristics

The Bohr Hydrogen Atom

Quantization, Bohr Radii, Electron Transitions, Atomic Spectra

Dual Nature of Electrons

Particle or Wave, Standing Wave, Particle Wave Length

The Schrödinger Wave Equation

Probability, Heisenberg Uncertainty Principle, Quantum Numbers

Quantum Numbers

Principal, Angular Momentum, Magnetic, and Spin Quantum Numbers, Orbitals

Multi-Electron Atoms

Orbitals, Shells, Subshells, Aufbau Principle, Pauli Exclusion Principle, Hund's Rule, Electron Configuration, Paramagnetic vs. Diamagnetic

Periodicity and Chemical Trends

Ionization Energy, Electron Affinity, Electronegativity, Atomic and Ionic Sizes, Oxidation States

3.A. Vocabulary

Write your definition of each term in the space provided.

Amplitude

Wavelength

Frequency

Speed

Photon

Ground State

Excited State

Phases of Waves

Node

Intensity of Waves

Wave Function

Quantum Number

Principle Quantum Number

Angular Momentum Quantum Number

Magnetic Quantum Number

Spin Quantum Number

Orbital

Radial Node

Angular Node

Subshell

Shell

Valence Electron

Electron Configuration

Paramagnetic

Diamagnetic

First Ionization Energy

Electron Affinity

Electronegativity

3.B.   Concepts

Write your notes on each concept in the space provided.

The Nature of Light–Wave or Particle?

Four Problems with the Wave Theory

The Bohr Model of the Atom

Particles as Waves

Electrons as Standing Waves around the Nucleus

Schrödinger Wave Equation

Heisenberg Uncertainty Principle

Types of Quantum Numbers

Types of Orbitals and their Characteristics

Orbitals in Multi-Electron Atoms

Pauli Exclusion Principle and Electron Spin

Hund's Rule

Writing Electron Configurations

Determining if an Atom is Paramagnetic or Diamagnetic

Periodic Trends in:

    (a) First Ionization Energy

    (b) Electron Affinity

    (c) Electronegativity

(d) Atomic Radius

(e) Ionic Sizes

(f) Oxidation States

3.C.   Problem Solving

Write your answer to each problem in the space provided. Try the problem yourself before reading the answer in the Tutorial.

3.1.   A two ton African elephant is running straight for you with a velocity of 20 mph. You hope (in vain) that the elephant's wave nature will take over before he hits you. What is the wavelength of the elephant?

3.2.   Why can an electron not be confined to a nucleus of radius 1×10^{-15} m?

3.3.   a)   Which principle quantum level can have $l = 0$, 1, and 2?

b)   Which l value can have $m = -2, -1, 0, 1$, and 2 only?

3.4.   a) What does 3d mean (n, l, and m)?

b) What does 4f mean (n, l, and m)?

c) What does 2s mean (n, l, and m)?

3.5.   What is the electron configuration of lithium?

3.6. What is the electron configuration of carbon?

3.7. What is the electron configuration for aluminum?

3.D.   Other Notes

CHAPTER FOUR

Bonding and Structure

Bonding

Bond Order, Percent Ionic Character, Polarity, Dipole Moment

Shapes of Molecules

Complex Ions, Lewis Dot Structures, VSEPR Theory, Geometries

Shapes of Molecules with Lone Pairs

Bond Angle, Geometry of Molecule vs. Geometry of Electrons

Valence Bond Theory

Introduction, Bond Types

Hybridization

Molecular Geometry

Molecular Orbital Theory (parts 1, 2, and 3) and Spectroscopy

Bond Order, Energy Correlation Diagrams, MO Types, Homonuclear Diatomics, Valence Bond vs Molecular Orbital Theory, HOMO, LUMO, Atmospheric Photochemistry, Ozone Layer

4.A.   Vocabulary

Write your definition of each term in the space provided.

Bond

Polar Bond

Lone Pair

Bonding Pair

Steric Number

Sigma (σ) Bonds

Pi (π) Bonds

Hybridization

Diamagnetic

Paramagnetic

HOMO

LUMO

4.B.   Concepts

Write your notes on each concept in the space provided.

Relationship of Bond Order, Length, and Energy

Percent Ionic Character and Dipole Moment

Polar *vs.* Nonpolar Molecules

Using VSEPR to Predict Molecular Shapes

Geometry of Electron Pairs *vs.* Geometry of Molecule

Valence Bond Theory

Sigma (σ) Bonding

Pi (π) Bonding

Hybrid Orbitals

Hybridization in Molecules with Double and Triple Bonds

Molecular Orbital (MO) Theory

Combining AOs to Generate MOs

H_2 Molecular Orbitals

Bond Orders from MO Theory

Characteristics of σ *vs.* π type MOs

Criteria which allows AOs to mix to form MOs

Energy Level Correlation Diagrams

    (a) for Homonuclear Diatomics

(b) for Heteronuclear Diatomics

Delocalized π Molecular Orbitals

HOMO and LUMO

Changes in Electronic, Vibrational and Rotational States

Atmospheric Photochemistry

4.C.   Problem Solving

Write your answer to each problem in the space provided. Try the problem yourself before reading the answer in the Tutorial.

4.1.   What is the wavelength of light needed to break a C=C bond given that the bond enthalpy is 615 kJ/mol.

4.2.   Draw the Lewis dot structure of each molecule in preparation for using VSEPR to determine its geometry. Determine the steric number of each.

HCN

BH_3

CH_4

PF_5

SF_6

4.3. Determine the Lewis dot structure, steric number, three-dimensional structure, and polarity of CO_2.

4.4. What would you predict is the three-dimensional structure of the following?

XeF_4

I_3^-

4.5. What is the hybridization on the central atom in:

a) CCl_4

b) H_2S

c) XeF_4

4.6. Draw the energy correlation diagram for Be_2.

4.D.   Other Notes

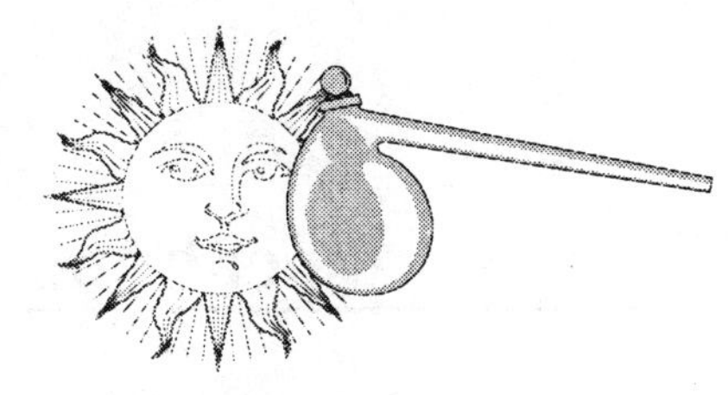

CHAPTER FIVE

Gases

Introduction

States of Matter, Air, Gas Characteristics, Pressure, Units, Temperature

Ideal Gas Equation

Gas Constant, Ideality, Charles', Boyle's, and Avogadro's Laws

Application of the Ideal Gas Equation

Ideal Assumptions, STP, Molar Volume, van der Waals Equation

Gas Mixtures

Partial Pressure, Dalton's Law

Kinetic Molecular Theory

Assumptions, Temperature, Absolute Zero

Molecular Motion and Temperature

Kinetic Energy, Root Mean Square Velocity, Maxwell-Boltzmann Plot

Pressure and Molecular Collisions

Pressure, Effusion, Diffusion

5.A. Vocabulary

Write your definition of each term in the space provided.

Solid

Liquid

Gas

Pressure

Barometer

Atmosphere

STP

Partial Pressure

Temperature

Absolute Zero

Root Mean Square (rms) Velocity

Effusion

Diffusion

Mean Free Path

5.B. Concepts

Write your notes on each concept in the space provided.

States of Matter

Pressure and Temperature

Ideal Gas Equation

Charles', Boyle's and Avogadro's Laws

The van der Waals Equation

Dalton's Law

Kinetic Molecular Theory

Absolute Zero

Root Mean Square (rms) Velocity

Effusion and Diffusion

5.C.　　Problem Solving

Write your answer to each problem in the space provided. Try the problem yourself before reading the answer in the Tutorial.

5.1.　Convert 7.54 kPa to:

atm

mmHg

in. H_2O

5.2.　Calculate the volume of 1.00 mole of an ideal gas, assuming it is at STP.

5.3. A 1.00 L sample of gas at STP is heated to 200.0 °C at constant volume. What is the final pressure of the gas?

5.4. A sample of gas at 100.0 °C in a cylinder 1.00 ft. long and 6.00 in. in diameter contains 1.68 g nitrogen gas and some neon gas. If the total pressure in the cylinder is 0.500 atm, how many moles of neon are present?

5.5. Determine the ratio of rates of diffusion for two gases (H_2 and Ne) at the same temperature.

5.D. Other Notes

CHAPTER SIX

Phase Changes and Solutions

Introduction

Phases, Vapor Pressure, Equilibrium

Phase Changes

Types, Critical and Triple Points, Sublimation, Cooling and Warming Curves

Solutions

Types, Colloids, Brownian Motion, Concentration Units, Making Solutions

Raoult's Law

Colligative Properties, Solvent Vapor Pressure, Deviations

Henry's Law

Volatile Solutions, Solute Vapor Pressure

Boiling Point and Freezing Point Changes

Boiling Point Elevation, Freezing Point Depression

Osmosis and Distillation

Solvent Transfer, Applications, Azeotropes

6.A.　　Vocabulary

Write your definition of each term in the space provided.

Phases of Matter

Vapor Pressure

Equilibrium

Phase Transition

Phase Diagram

Equilibrium Line

Critical Point

Supercritical Fluid

Triple Point

Sublimation

Solution

Colloid

Solvent

Solute

Weight Percent (Mass Percent)

Mole Fraction

Molarity

Molality

Colligative Properties

Normal Boiling Point

Osmosis

Distillation

Azeotrope

6.B.　　Concepts

Write your notes on each concept in the space provided.

Vapor pressure above a liquid

Features of Phase Diagrams

Phase Changes (or Transitions)

Cooling and Warming Curves

Types of Solutions

Concentration Units and Conversions

Making Solutions

Vapor Pressure Lowering

Raoult's Law

Henry's Law

Boiling Point Elevation

Freezing Point Depression

Osmotic Pressure

Reverse Osmosis

Write your answer to each problem in the space provided. Try the problem yourself before reading the answer in the Tutorial.

6.1. A 1.00 *M* solution of some substance (MW = 130.0 g/mol) has a density of 1.05 g/mL. What is the weight percent and molality of the substance in this solution?

6.2. 244 g of pentanol are added to 49.8 g of water at 25°C. What will be the vapor pressure of the water above this solution?

6.3. A solution was made by adding 10.00 g butanol to 100.0 g water at 25°C. Pure water has a vapor pressure of 23.76 torr at this T. The observed vapor pressure was 23.20 torr. What is the molar mass of butanol?

6.4. If 0.0500 g of O_2 are dissolved in 600.0 g of water, what is the vapor pressure of the oxygen at 25°C? Note: $k(O_2) = 4.34 \times 10^4$.

6.5. While cooking spaghetti, you put a large pinch of salt (115 grams) into 1.00 liter of water. By how much did you raise the boiling point? The k_b for water is 0.51 kg K/mol.

6.6. A chemist dissolves 5.00 grams of an unknown into 1.00 kg water, thereby lowering the freezing point by 0.0517°C. She knows that the empirical formula of the compound is CH_2O and that the compound does not dissociate or ionize in water. What is the molecular formula of the unknown? The k_f (H_2O) = 1.86 kg°C/mol.

6.D. Other Notes

Thermodynamics

Introduction

Vocabulary, Thermodynamic States, Reversibility, State Functions

First Law of Thermodynamics

Conservation of Energy, P-V Work, Sign Conventions

Heat

Specific Heat, Heat Capacity, Calorimetry, Constant Volume and Pressure Calculations

Thermochemistry

Enthalpy, Exothermic and Endothermic Processes, Hess' Law, Standard Enthalpy, Isobaric and Isochoric Processes

Bond Enthalpy

Calculating Reaction Enthalpy Changes

Reversible Processes

Isothermal and Adiabatic Processes, Carnot Cycle

7.A.   Vocabulary

Write your definition of each term in the space provided.

Thermodynamics

System

Open System

Closed System

Surroundings

Intensive Property

Extensive Property

Thermodynamic State

Thermodynamic Process

Reversible Process

State Function

Work

Heat

Calorie

Heat Capacity

Calorimeter

Thermochemistry

Enthalpy

Exothermic

Endothermic

Bond Enthalpy

Isochoric

Isobaric

Isothermal

Adiabatic

Efficiency

Isotherm

7.B. Concepts

Write your notes on each concept in the space provided.

First Law of Thermodynamics

P-V Work

Calculating Heat

Specific Heat and Heat Capacity

Using Bomb Calorimeters

Enthalpy and Energy

Endothermic *vs.* Exothermic Processes

Hess' Law

Standard Enthalpy Changes

Using Standard Enthalpies of Formation

Using Bond Enthalpies

Types of Reversible Processes

Carnot Cycle

*Write your answer to each problem in the space provided. Try the problem
yourself before reading the answer in the Tutorial.*

7.1.   A gas has a volume of 5.00 L. It is compressed to a new volume of 1.00 L using a
pressure of 40.0 atm. Calculate the work done on the gas.

7.2.   1.0 g of TNT ($C_7H_5N_3O_6$) is combusted in a bomb calorimeter. The heat capacity
of the calorimeter is 10.0 kJ/°C. The temperature of the water inside the
calorimeter is raised 1.51°C. What is the heat of combustion of TNT?

7.3. Propane is combusted at a constant temperature of 25°C. If 1.25 mol O_2 are completely combusted with propane, how much heat is given off? Important items: Assume constant atmospheric pressure. The standard heat (enthalpy) of combustion of propane at 25°C is -2.22×10^6 J/mol.

7.4. What is $\Delta H°$ for $2\ NO_{(g)} + O_{2\ (g)} = 2\ NO_{2\ (g)}$? Use heats of formation.

7.5. What is the $\Delta H°$ for the incomplete combustion of 1.0 mole of hexanol? Assume a 2:1 ratio of CO_2 to CO produced.

7.D. Other Notes

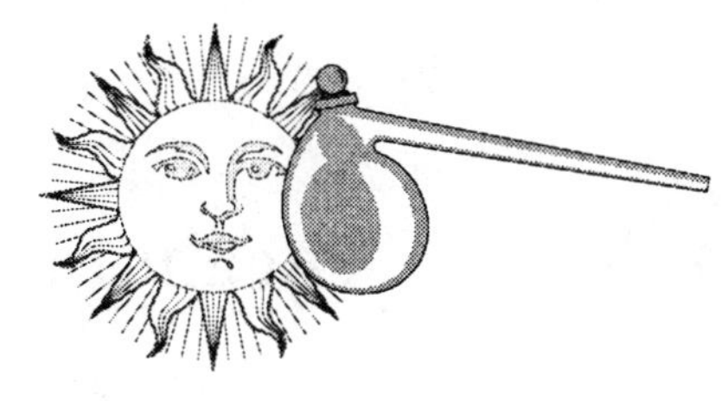

CHAPTER EIGHT

Spontaneous Change and Equilibrium

Entropy

Reversible Processes, Driving Forces, 2nd Law of Thermodynamics, System vs. Surroundings

Calculating Entropy

Adiabatic, Isothermal, and Isochoric Processes, Phase Changes, Trouton's Rule

Statistical Thermodynamics

Probability, Ordered Systems, Spontaneity, Microstates, Boltzmann's Relationship and Constant

Third Law of Thermodynamics

Absolute Entropies, Absolute Zero, Standard Molar Entropies

Gibbs Free Energy

Spontaneity, Derivation, Driving Forces, Equilibrium, Dissolution

ΔG°_f

Free Energies of Formation, Reaction Spontaneity

8.A. Vocabulary

Write your definition of each term in the space provided.

Entropy

Isentropic

Statistical Thermodynamics

Microstate

Standard Molar Entropy

8.B. Concepts

Write your notes on each concept in the space provided.

Calculating Entropy Changes

Second Law of Thermodynamics

Driving Forces in Nature

ΔS of Phase Changes

ΔS for Isothermal Gas Expansion

ΔS for Isochoric Gas Heating

Randomness and Entropy

Microstates and Probability

Boltzmann's Relationship

Third Law of Thermodynamics

Determining Standard Molar Entropies

Determining ΔS reaction from $S°$ values

Free Energy and Spontaneity

ΔG and Nature's Driving Forces

Using Free Energies of Formation

8.C.   Problem Solving

Write your answer to each problem in the space provided. Try the problem yourself before reading the answer in the Tutorial.

8.1.   A gas is heated from a temperature of 25°C to 125°C, yet no P-V work is done. What is the ΔE for this process if the ΔS is 475 J/K?

8.2.   One mole of gas shrinks to one-third its original volume at constant T. What is the ΔS for this process?

8.3. What is the ΔS° for the reaction $N_{2\,(g)} + 2\,O_{2\,(g)} = 2\,NO_{2\,(g)}$?
$[S^\circ\,(O_{2\,(g)}) = 205.03\ \text{J/molK}]$
$[S^\circ\,(N_{2\,(g)}) = 191.50\ \text{J/molK}]$
$[S^\circ\,(NO_{2\,(g)}) = 239.95\ \text{J/molK}]$

8.4. Tungsten (VI) oxide can be reduced to tungsten by the following reaction: $2\,WO_{3\,(s)} + 3\,C_{(s)} = 2\,W_{(s)} + 3\,CO_{2\,(g)}$. Is the reaction feasible (spontaneous) at room temperature? If not, what could be done to make it spontaneous?

8.D. Other Notes

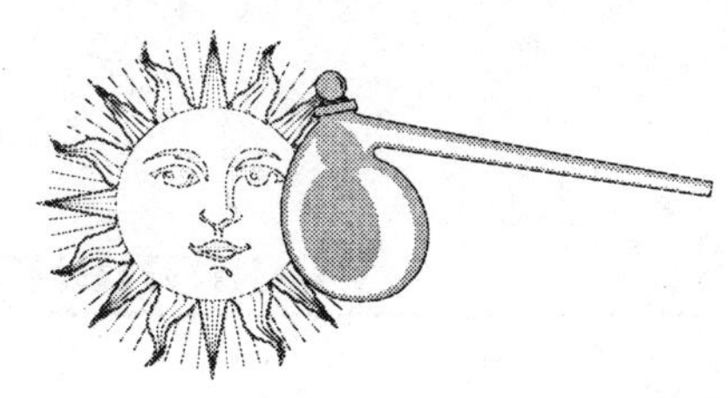

CHAPTER NINE

Chemical Equilibrium

Introduction

Equilibrium Constants, Law of Mass Action, Activity

Using Equilibrium Constants

Example Problems, Approximations

Manipulating K

Adding/Reversing/Multiplying Equations by a Constant, Temperature Dependence, Concentration-Based Constants

Reaction Quotient

Kinetic Factors, Le Châtelier's Principle, Temperature Dependence

ΔG Relates to K

Derivation, Thermodynamics vs. Kinetics

ΔG not at Equilibrium

Reaction Quotient, Activity, Le Châtelier's Principle vs. van't Hoff Equation

Separations

Partition Coefficient, Solvent Extraction, Miscibility, Paper and Column Chromatography

9.A.   Vocabulary

Write your definition of each term in the space provided.

Equilibrium

Dynamic

Equilibrium Constant

Activity

Successive Approximations

Reaction Quotient

Immiscible

Extractant

Chromatography

Mobile Phase

Stationary Phase

9.B.   Concepts

Write your notes on each concept in the space provided.

The Nature of Equilibrium

Equilibrium Constant Expressions

Activity and Units of K

Using Successive Approximations

Deriving Values of K when:

    (1) we add chemical reactions

    (2) we reverse the chemical reaction

    (3) we multiply the chemical reaction by a constant

Temperature Dependence of K

K's based on pressure *vs.* concentration

Using the Reaction Quotient

Le Châtelier's Principle

Partition Coefficients

Solvent Extraction Separations

Chromatographic Separations

9.C.   Problem Solving

Write your answer to each problem in the space provided. Try the problem yourself before reading the answer in the Tutorial.

9.1.   If we place 10.0 g of NO_2 in a 2.00 L bottle at 25 °C, what will be the number of moles of NO_2 at equilibrium? Use the equation: $2\ NO_{2\ (g)} = N_2O_{4\ (g)}$. $K_p = 8.8$ (25°C).

9.2.   If we mix N_2 at 10.00 atm with H_2 also at 10.00 atm in a 1.00 L container, what will be the pressure of H_2 at equilibrium? Use the equation: $N_2 + 3\ H_2 = 2\ NH_3$. $K = 6.8 \times 10^5$ at 25°C.

9.3. $N_{2\,(g)} + 3\,H_{2\,(g)} = 2\,NH_{3\,(g)}$. Write the expressions for K_p and K_c.

9.4. $CH_{4\,(g)} + H_2O_{(g)} = CO_{(g)} + 3\,H_{2\,(g)}$. $K_p = 1.8 \times 10^{-7}$ at 600 K. If we mix 3 atm each of CH_4 and of H_2O with 2 atm each of CO and of H_2, which way will the reaction proceed to reach equilibrium? Use Q to answer this question.

9.5. Calculate K for this reaction from the value of $\Delta G°$: Ni $_{(s)}$ + 2 HCl $_{(g)}$ = NiCl$_2$ $_{(s)}$ + H$_2$ $_{(g)}$. Hint: Use the following $\Delta G°_f$ values to find $\Delta G°$. Use this to find K from the formula we just derived.

$\Delta G°_f$: [Ni $_{(s)}$ = 0 kJ/mol]

[NiCl $_{(s)}$ = −272 kJ/mol]

[H$_2$ $_{(g)}$ = 0 kJ/mol]

[HCl $_{(g)}$ = −95 kJ/mol]

9.6. During an experiment using this reaction: SO$_2$ + 2 H$_2$O = H$_3$O$^+$ + HSO$_3^-$, the following concentrations were measured: 0.10 M SO$_2$, 3.0 M H$_3$O$^+$, 3.0 M HSO$_3^-$. What is the ΔG for this reaction under these conditions? K = 1.5 x 10^{-2}.

9.7. Use the van't Hoff equation to calculate the ΔH° for the autoionization of water. The pK_w value changes with temperature as follows: $pK_w\,(0^\circ C) = 14.94$. $pK_w\,(60^\circ C) = 13.02$.

9.D.   Other Notes

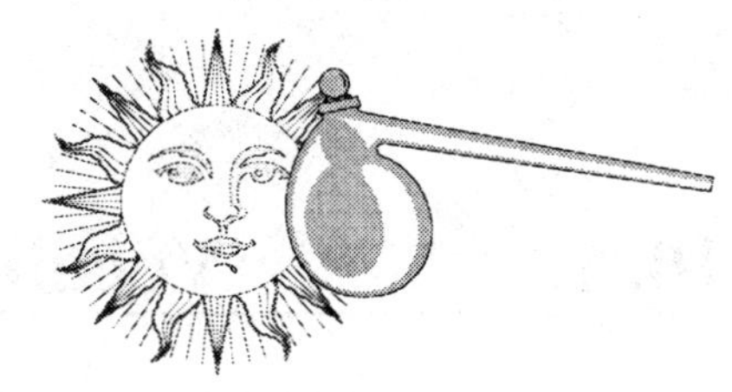

CHAPTER TEN

Acids and Bases

Introduction

Definitions, Conjugate Acids and Bases, Relative Strengths, Nomenclature

Acid-Base Equilibrium

Amphoteric Substances, Autoionization of Water, K_w

pH

pH Scale, Common pHs, Measuring pH, Indicators

Acid-Base Strength

Electrolytes, Weak vs. Strong, K_a , pK_a

K_a Relates to K_b

Hydrolysis, Acid vs. Conjugate Base Strengths

Buffers

Components, Buffer Capacity, Calculations

Titrations

Procedures, Different Types, Titration Curves, Indicators

Polyprotic Acids

Species Distribution Plots, Multiple Equilibria

10.A.   Vocabulary

Write your definition of each term in the space provided.

Acid

Base

Ionization

Conjugate Base

Amphoteric

pH Scale

Indicator

Electrolytes

Hydrolysis

Buffer

Titration

Titrant

Titrate

Equivalence Point

Titration Curve

Polyprotic Acid

Species Distribution Plot

Write your notes on each concept in the space provided.

Types of Acids

Types of Bases

Ionization of Acids

Conjugate Acids and Bases

Naming Acids

Auto-Ionization of Water and K_w

The pH scale

pH Indicators and Sensors

Strong and Weak Electrolytes

Strong and Weak Acids

K_a Values

Hydrolysis Reactions and K_h

How Buffers Work

Acid-Base Titrations Curves

Polyprotic Acid Chemistry

Write your answer to each problem in the space provided. Try the problem yourself before reading the answer in the Tutorial.

10.1. At 25°C, the K_a for hydrofluoric acid is 6.6 x 10^{-4}. What is the pK_a of HF? Does a higher pK_a mean a weaker or stronger acid?

10.2. For HF $K_a = 6.6$ x 10^{-4}. What is the pH of a 1.0 M solution of HF?

10.3. The sulfate ion is the conjugate base of the acid called the bisulfate ion, HSO_4^-. What is the pH of a 0.100 M. SO_4^{-2} solution? You can assume that the HSO_4^- ion does not in turn hydrolyze after it is formed because H_2SO_4 is a very strong acid. Use the equation: $HSO_4^- + H_2O = SO_4^{-2} + H_3O^+$. $pK_a = 1.92$.

10.4. What is the pH of an equimolar (0.100 M) acetate buffer? In other words, this is a buffer made 0.100 M in acetic acid and 0.100 M soluble acetate salt (e.g. sodium acetate). K_a for acetic acid is 1.76×10^{-5}.

10.5. Using this equimolar (0.100 M) acetate buffer, how much will the pH change if 0.010 mole of HCl is added to one liter of the buffer solution? K_a for acetic acid is 1.76×10^{-5}.

10.6. If you wanted to make your buffer 0.050 M HF, what concentration of NaF should you use to get a pH of 3.5?

10.7. 20.0 mL of 0.100 M $HC_2H_3O_2$ are titrated with 0.500 M NaOH. What is the pH after 4.00 mL NaOH are added to the acetic acid? Here is something you might need: K_a $(HC_2H_3O_2) = 1.8 \times 10^{-5}$.

10.8. If we know that $K_{a1} = 4.3 \times 10^{-7}$ and $K_{a2} = 4.8 \times 10^{-11}$, what is the pH of a 0.010 M carbonic acid solution? What is the equilibrium concentration of CO_3^{-2}?

10.D.   Other Notes

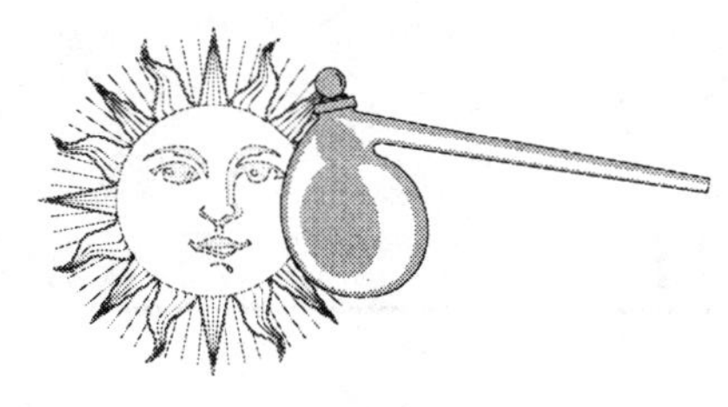

CHAPTER ELEVEN

Solubility Equilibria

Solubility

Solubility Rules, Increasing and Decreasing Solubility, Solvation

The Solubility Product, K_{sp}

Calculations, Temperature Effects

Shifting Equilibrium

Common Ion Effect, pH Effect

Complexation Effects

Complexation Terms, Instability and Stability Constants, Hydrated Metal Ions

11.A. Vocabulary

Write your definition of each term in the space provided.

Dissolution

Precipitation

Saturated

Unsaturated

Supersaturated

Solubility

Soluble

Insoluble

Slightly Soluble

Solubility Product

Common Ion Effect

Instability Constant

Stability Constant

11.B.   Concepts

Write your notes on each concept in the space provided.

Variation of Solubility with Temperature

Thermodynamic Driving Forces and Solubility

Solubility Rules

Using the Solubility Product

Effect of Common Ions on Solubility

Effect of pH on Solubility

Complexation Reactions

Effect of Complexation on Solubility

Write your answer to each problem in the space provided. Try the problem yourself before reading the answer in the Tutorial.

11.1.   Are these compounds water soluble?

$AgNO_3$

$CaCO_3$

KBr

$PbSO_4$

11.2.   Find the solubility of AgCl in water. $K_{sp} = 1.6 \times 10^{-10}$.

11.3. If I mix 100.0 mL of 0.100 M lead nitrate with 100.0 mL of 0.010 M sodium iodide, what will be the iodide concentration at equilibrium? K_{sp} of $PbI_{2(s)} = 1.4 \times 10^{-8}$.

11.4. If 0.100 moles of $NaIO_3$ are dissolved in water to make 1.00 L of solution, how much will the solubility of the lead(II) iodate decrease? $Pb(IO_3)_{2\ (s)} = Pb^{+2}_{\ (aq)} + 2\ IO_3^-{}_{(aq)}$. $K_{sp} = 2.6 \times 10^{-13}$.

11.D. Other Notes

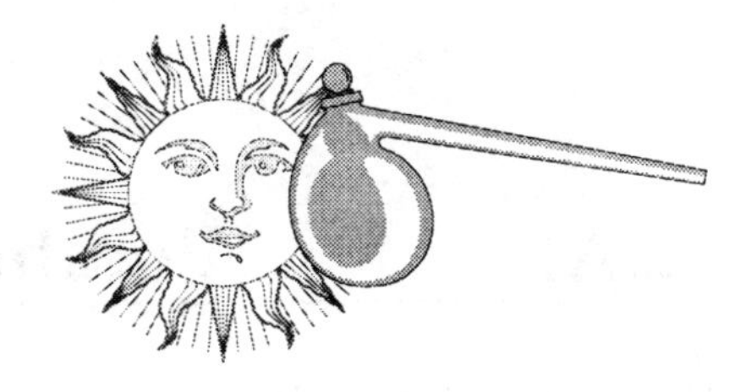

CHAPTER TWELVE

Electrochemistry

Introduction

Redox Reactions, Balancing, Half Reactions, Disproportionation

Electrochemical Cells

Components, Galvanic Cell, Cell Potential (Voltage), Δε, Electrolytic Cell

Electron Flow

Current, Faraday Constant, Electrolysis

Δε°

Standard Conditions, Calculations, Reduction Potentials

ΔG Relates to Δε

Adding Reactions, Predicting Δε°, Reduction Potential Diagrams, Water Fuel Cell, Electrolysis of Water

Nernst Equation

Example Problems

12.A.   Vocabulary

Write your definition of each term in the space provided.

Oxidation-Reduction Reaction

Disproportionation

Electrochemical Cell

Galvanic Cell

Voltage

Electrolytic Cell

Current

12.B.   Concepts

Write your notes on each concept in the space provided.

Balancing Oxidation-Reduction Reactions

Galvanic Cell Operation and Design

Factors which Determine $\Delta\varepsilon$

Electrolytic Cell Operation

Current and Voltage in Cells / Faraday's Constant

Using Standard Half-Cell Potentials

Relationship of ΔG to $\Delta \varepsilon$

Predicting New Half-Cell Potentials

Using Reduction Potential Diagrams

Electrolysis of Water

Nernst Equation

Write your answer to each problem in the space provided. Try the problem yourself before reading the answer in the Tutorial.

12.1. Identify the reduced species (oxidizing agent) and oxidized species (reducing agent) in this equation: $Br_{2\ (l)} + SO_{2\ (g)} = Br^-_{\ (aq)} + HSO_4^-$ (in acid). Balance this equation.

12.2. Balance this equation: $Cl_{2\ (g)} = ClO_3^-_{\ (aq)} + Cl^-_{\ (aq)}$ (in base).

12.3. In an electrolytic cell, Cu is deposited at one electrode from Cu^{+2} ions in solution. How many grams of Cu will be deposited each hour if the current is 1.00 A?

12.4. $\Delta\varepsilon$ (measured) $= 0.963$ V at $25°C$ for the following cell: $Cu_{(s)} / Cu^{+2}_{(aq)} (1.00\ M)\ \|\ Br_{2\,(l)} / Br^-_{(aq)}$. What is the bromide concentration?

$$Cu^{+2}_{(aq)} + 2\ e^- = Cu_{(s)} \qquad E° = 0.340\ V$$
$$Br_{2\,(l)} + 2\ e^- = 2\ Br^-_{(aq)} \qquad E° = 1.065\ V$$

12.D.   Other Notes

CHAPTER THIRTEEN

Kinetics

Introduction

Kinetics vs. Thermodynamics, Radiocarbon Dating, Reaction Rate Laws, Zero and First Order Processes, Half-life

Rate Laws

First and Second Order Processes, Rate Data and Rate Laws

Mechanisms

Elementary Steps, Molecularity, Relation Between Kinetics and Thermodynamics, Rate Determining Step

Transition State Theory

Temperature Effects on Rate, Boltzmann Diagram, Activation Energy, Activated Complex, Potential Energy Diagrams, Endothermic vs. Exothermic Reactions

Catalysts

Uncatalyzed vs. Catalyzed Pathways, Activation Energy, Classes of Catalysts, Enzymes

Write your definition of each term in the space provided.

Kinetics

Reaction Rate

Half-Life

Elementary Step

Molecularity

Rate Determining Step

Activation Energy

Transition State

Catalyst

Homogeneous Catalyst

Heterogeneous Catalyst

Enzyme

13.B. Concepts

Write your notes on each concept in the space provided.

Kinetics *vs.* Thermodynamics

Reaction Rate Expressions

Zero Order Kinetics

First Order Kinetics

Using Half-Life

Second Order Kinetics

Reaction Mechanisms and Elementary Steps

Relations Between Rate Constants and Equilibrium Constants

Effect of Temperature on Reaction Rate

Transition State Theory

How Catalysts Work

Write your answer to each problem in the space provided. Try the problem yourself before reading the answer in the Tutorial.

13.1. SO_2Cl_2 at 1.00 atm is placed in a closed vessel at 320°C. After 29.1 hours the pressure of SO_2Cl_2 has dropped to 0.100 atm as a result of the first order decomposition reaction: $SO_2Cl_2{}_{(g)} = SO_2{}_{(g)} + Cl_2{}_{(g)}$. What is the first order rate constant for this reaction?

13.2. Consider the reaction: A + B = AB. The rate of synthesis of AB is found to be first order in B. If we start with 3.2 M B, what will be the concentration of B after 5 minutes? $t_{1/2} = 35.0$ s.

13.3. Describe the kinetics by finding the rate laws of the decomposition of H_2O_2 given the following data:

Time (s)	$[H_2O_2]$ (M)
0.0	1.500
90.0	1.392
180.0	1.292
270.0	1.199
360.0	1.113
450.0	1.032
540.0	0.958
630.0	0.889
720.0	0.823
810.0	0.766
900.0	0.711

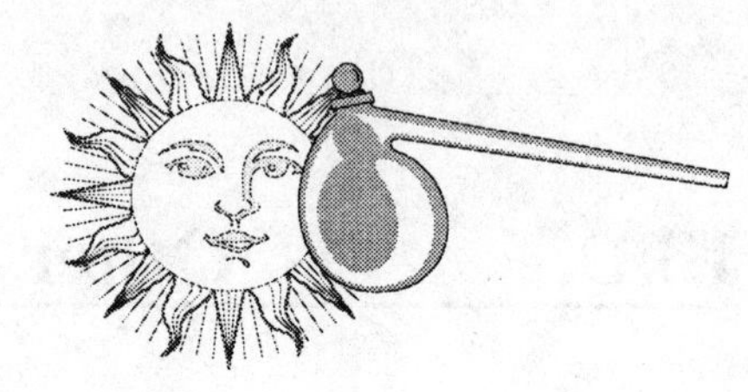

CHAPTER FOURTEEN

Nuclear Chemistry

Introduction

The Nucleus, Isotopes, E = mc², Binding Energy

Nuclear Decay

Radioactivity, Antiparticles, α, β, γ, and Positron Emission, Electron Capture, Geiger Counter

Kinetics of Radioactive Decay

First Order Kinetics, Radioactive Dating, Carbon Dating

Uses of Nuclear Processes

Fission, Fusion, Nuclear Reactors, Breeder Reactors

 Vocabulary

Write your definition of each term in the space provided.

Atomic Mass Unit

Nuclide

Nucleon

Isotope

Binding Energy

Electron Volt (eV)

Million Electron Volt (MeV)

Radioactivity

Antiparticle

Positron

Antiproton

Neutrino

Antineutrino

Alpha (α) Decay

Beta (β) Decay

Gamma (γ) Ray

Radioactive Dating

Nuclear Fission

Breeder Reactor

Nuclear Fusion

14.B. Concepts

Write your notes on each concept in the space provided.

Nuclear Structure and Isotopes

Determining Nuclear Binding Energy

Fundamental Particles

Types of Nuclear Decay

How a Geiger Counter Works

Kinetics of Radioactive Decay

Radioactive Dating

Nuclear Fission and Chain Reaction

Nuclear Fusion

14.C. Problem Solving

*Write your answer to each problem in the space provided. Try the problem
yourself before reading the answer in the Tutorial.*

14.1. What is the nuclear binding energy per mole of the hydrogen isotope called
tritium ($^3_1$H)?
m ($^1_1$p) = 1.00727647 amu
m ($^1_0$n) = 1.00866490 amu
m ($^0_{-1}$e) = 5.4857990 x 10^{-4} amu
m ($^3_1$H) = 3.016049 amu

14.2. Write the nuclear decay equations in the $^{238}_{92}$U decay series going to $^{206}_{82}$Pb.

14.3. A new computer program comes out onto the market. It is so good that everyone in the world (assume 5.0 billion strong) just *has* to have his/her own copy. After 3.0 days, half of the world population has this program. At this rate, how much longer will it be until 99 % of the population has this program, assuming that the rate of sales is described by first order kinetics.

14.4. The isotope ^{87}Rb decays to ^{87}Sr with a half-life of 4.9 x 10^{10} yr. How old is a rock that contains 1.0 ^{87}Sr atom for every 120.0 ^{87}Rb atoms?

14.D. Other Notes

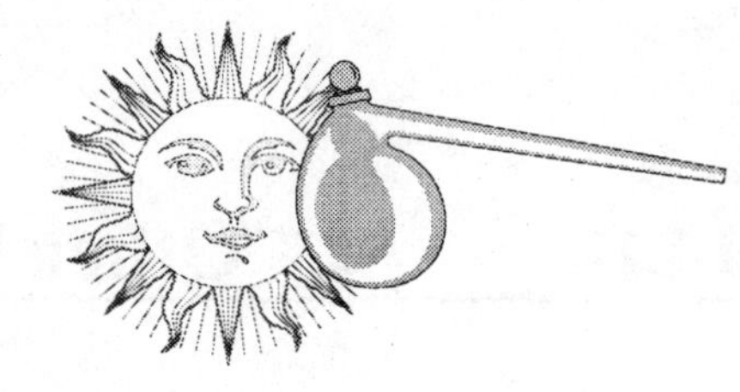

Coordination Chemistry

Metal Ion Complexation

Aqueous Ions, Complex Formation, Nomenclature

Shapes of Coordination Complexes

Geometry, Coordination Number, Geometric Isomers

Optical Isomerism

Chirality, Enantiomers, Tetrahedral and Octahedral Structures

Crystal Field Theory

Colors of Complexes, d-Splitting for Various Geometries, Δ, Spectrochemical Series, Low and High Spin

Ligand Field Theory and Interesting Complexes

MO Diagram, Pi Back-Bonding, Organometallic Complexes, Macrocyclic Complexes

Write your definition of each term in the space provided.

Ligand

Monodentate Ligand

Bidentate Ligand

Coordination Sphere

Coordination Number

Chelate

Optical Isomer

Enantiomer

Diastereomer

Chirality

Chiral Center

Organometallic Compound

Macrocyclic Ligand

Write your notes on each concept in the space provided.

Structure of the Coordination Sphere

Bonding in Coordination Complexes

Naming Coordination Complexes

Geometries with Various Coordination Numbers

Geometries and Optical Isomerism

Crystal Field Splitting of d-Orbital Energies

Low Spin *vs.* High Spin Complexes

Ligand Field Theory

Pi Back-Bonding

Write your answer to each problem in the space provided. Try the problem yourself before reading the answer in the Tutorial.

15.1.   Name the following complexes:

     a) $[Ni(H_2O)_4(OH)_2]$

     b) $[HgClI]$

     c) $K_3[Co(CN)_6]$

     d) $[FeBrCl(en)_2]Cl$

15.2.   Determine the coordination number, systematic name or formula, and shape for the following. Do any geometric isomers exist?

     a) $[Pt(NH_3)_2Cl_2]$

b)	$K_2[Co(NH_3)(CN)_5]$

c)	$[CuCl_2]^-$

d)	Dichlorodihydroxozincate(II) ion

e)   Pentaaminecarbonatoiron(III) chloride

15.3. K^+ complexes with 18-crown-6 in an equilibrium reaction. If a chemist makes a solution 50.00 mM KCl and 50.00 mM 18-crown-6, after a while she finds the concentration of free K^+ ion in solution is 0.0600 mM. Determine the value of the formation constant for this complex.

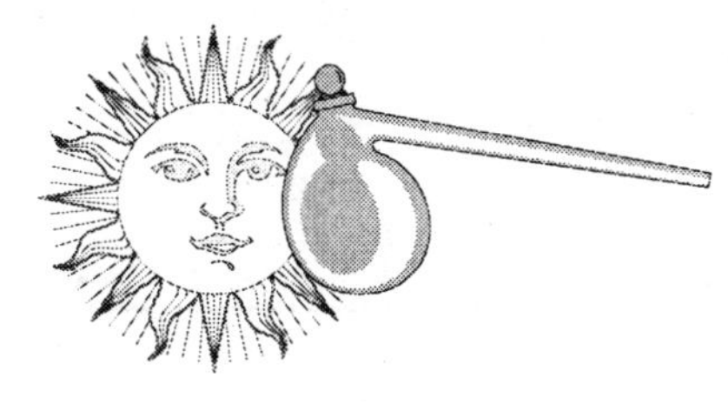

CHAPTER SIXTEEN

Intermolecular Forces and Solids

Attraction

Intermolecular Forces, Electrostatic Forces, Percent Ionic Character, Dispersion Forces, Hydrogen Bonding

Dissolution

Driving Forces in Dissolution, Lattice Energy, Hydration Energy, Heat of Solution, Entropy

Atomic Crystals and the Unit Cell

Stacking Arrangements, HCP, CCP, Unit Cell Characteristics, Size, Density

Ionic Crystals

Characteristic Lattice Holes, Carbon: Diamond and Graphite, X-Ray Diffraction, Metallic Crystals, Conducting Bands

Crystal Defects

Point Defects, Non-Stoichiometric Compounds, Amorphous Solids, Liquid Crystals

___

Write your definition of each term in the space provided.

Point Defect

Schottky Defect

Frenkel Defect

Vacancy Defect

F-Center

Non-Stoichiometric Compounds

Amorphous Solids

Glasses

Liquid Crystals

Electrostatic Forces

a) Ion-Ion

b) Ion-Dipole

c) Dipole-Dipole

Dipole

Percent Ionic Character

van der Waals (London) Force

Polarizable

Hydrogen Bond

Lattice Energy

Hydration Energy

Heat of Solution

Close Packing

Hexagonal Close (ABA) Packing

Cubic Close (ABC) Packing

Cubic Unit Cell:

a)   Simple

b)   Body-Centered

c)   Face-Centered

Constructive Interference

Bragg Equation

Band Gap

16.B.　　　　Concepts

Write your notes on each concept in the space provided.

Defects in Crystals

Dipole Trends

Lattice Energy Trends

Hydration Energy Trends

Atomic Stacking in Crystals

Atoms per Unit Cell

Size of Unit Cell

Holes in Unit Cell

X-Ray Diffraction

Metallic Crystals

Write your answer to each problem in the space provided. Try the problem yourself before reading the answer in the Tutorial.

16.1. Rank the following compounds according to increasing boiling point based on intermolecular forces:

H_2O

$CHCl_3$

H_2S

CH_3COOH

Br_2

Xe

16.2. Is LiCl spontaneously soluble in water? $\Delta H_{HYD} = -865.1$ kJ/mol. $\Delta H_{LATTICE} = -828$ kJ/mol.

16.3. Aluminum exhibits cubic close packing. If aluminum's atomic radius is 143.2 pm, what is the density of aluminum?

16.D.   Other Notes